Lia Colwell

Integration of Opuntia stricta into the Diet of Olive Baboons in Kenya

AF153158

Lia Colwell

Integration of Opuntia stricta into the Diet of Olive Baboons in Kenya

LAP LAMBERT Academic Publishing

Impressum / Imprint
Bibliografische Information der Deutschen Nationalbibliothek: Die Deutsche Nationalbibliothek verzeichnet diese Publikation in der Deutschen Nationalbibliografie; detaillierte bibliografische Daten sind im Internet über http://dnb.d-nb.de abrufbar.
Alle in diesem Buch genannten Marken und Produktnamen unterliegen warenzeichen-, marken- oder patentrechtlichem Schutz bzw. sind Warenzeichen oder eingetragene Warenzeichen der jeweiligen Inhaber. Die Wiedergabe von Marken, Produktnamen, Gebrauchsnamen, Handelsnamen, Warenbezeichnungen u.s.w. in diesem Werk berechtigt auch ohne besondere Kennzeichnung nicht zu der Annahme, dass solche Namen im Sinne der Warenzeichen- und Markenschutzgesetzgebung als frei zu betrachten wären und daher von jedermann benutzt werden dürften.

Bibliographic information published by the Deutsche Nationalbibliothek: The Deutsche Nationalbibliothek lists this publication in the Deutsche Nationalbibliografie; detailed bibliographic data are available in the Internet at http://dnb.d-nb.de.
Any brand names and product names mentioned in this book are subject to trademark, brand or patent protection and are trademarks or registered trademarks of their respective holders. The use of brand names, product names, common names, trade names, product descriptions etc. even without a particular marking in this work is in no way to be construed to mean that such names may be regarded as unrestricted in respect of trademark and brand protection legislation and could thus be used by anyone.

Coverbild / Cover image: www.ingimage.com

Verlag / Publisher:
LAP LAMBERT Academic Publishing
ist ein Imprint der / is a trademark of
OmniScriptum GmbH & Co. KG
Heinrich-Böcking-Str. 6-8, 66121 Saarbrücken, Deutschland / Germany
Email: info@lap-publishing.com

Herstellung: siehe letzte Seite /
Printed at: see last page
ISBN: 978-3-659-54837-6

ABSTRACT

Wild free ranging baboons in Mukugodo Division, Kenya have started in recent years to incorporate *Opuntia stricta* into their diets. Between August 2011 and February 2012 the contribution of Opuntia on foraging strategies was investigated for one habituated olive baboon troop. A breakdown of monthly feeding patterns was analyzed for variation in overall foraging behaviors and variation between adult male and female baboons. Food categories were grouped by vegetation type into Opuntia fruits, grass/herb species, bush/tree species, *Acacia* seeds and pods, and other succulents and miscellaneous items, which were defined as 'food types'. The mean percentage of time spent on individual food types was significantly different ($F_{4,285} = 70.43$, $p < 0.0001$). Grass and herb species were eaten the most overall (52% S.E.$\pm$1.87), followed by Opuntia fruits (25% $\pm$0.94). The rate of Opuntia fruits eaten per minute was significantly inversely correlated with grass and herb species ($r = -0.616$, $n = 63$, $p < 0.0001$), bush and tree species ($r = -0.299$, $n = 63$, $p = 0.017$), and positively correlated with succulents and miscellaneous species ($r = 0.314$, $n = 63$, $p = 0.012$). Males and females both preferred ripe fruits ($F_{2,167} = 10.157$, $p < 0.0001$). Males ate more Opuntia fruits than females ($t = 5.383$, $df = 25$, $p < 0.0001$) while females spent more time on herb layer ($t = 2.836$, $df = 25$, $p = 0.009$). This study adds to an understanding of the impact of new, energy-rich foods on the dynamics of diet and foraging behavior. It also contributes to how baboons construct foraging strategies as Opuntia continues to alter the landscape.

TABLE OF CONTENTS

LIST OF ACRONYMS

BRD - Bridal sleeping site

CRIP - Cripple Troop

MLK - Malaika Troop

NMU - Namu

OFT - Optimal Foraging Theory

PHG - Pumphouse Gang Troop

SIS - Sisal sleeping site

STT - Soitoitashe Troop

TWG - Twala Gully

UNBP - Uaso Ngiro Baboon Project

WTR - White Rocks sleeping site

ACKNOWLEDGEMENTS

I would like to thank Dr. Shirley Strum for allowing me the opportunity to work with the Uaso Ngiro Baboon Project (UNBP) and conduct this research. Her advice and guidance from project conception, fieldwork, analysis, and the writing stages, has been immeasurable. I would also like to thank UNBP for allowing me access to supplemental data which helped to fill out this project and provide necessary insights.

I would like to recognize Jes Graham and Rose Argall, the Deputy Directors of UNBP, for their technical advice and guidance throughout the course of my project. Thank you to Jes for helping me with the full day data samples and aiding in the creation of the data sheets.

Many thanks to all the UNBP field researchers and trackers for their assistance while out with the troop. I learned many valuable techniques from them in regards to field research and ecological data collection. Thank you to John Lendoyan for his Masaai stories and songs, and particularly positive attitude during that long ride to Nanyuki Cottage Hospital.

Thank you to Dr. Kimanzi for his tireless help during the analysis and correction process, a process made a million times smoother due to his vigilant supervision. Thank you also to Prof. Wahungu for his advice and supervision throughout.

CHAPTER ONE

Introduction

1.1 Background Information

Fruits are a great source of energy, and Opuntia fruits are no exception. Ripe Opuntia fruits are typically softer, easier to manipulate, more palatable, and generally contain more calories than unripe fruits. I predicted that ripe Opuntia fruits will be harvested more by baboons than unripe fruits. When the fruits are eaten over the course of a day can determine if they are used to fill up fast in order to make time for other activities, or treated as a food that can supplement the baboons' diet when extra calories are needed, thus detailing a portion of the baboons' foraging style.

Variation in the diets of adult male and female baboons can be predicted due to differences in body size and energy needs (Strum, 2009). Males require more energy than females due to their larger body size, and their larger gut means they can eat foods with less nutritional quality (Altmann and Alberts, 1987; Strum, 1991). Reproduction and consortships can also impact foraging behaviors. Mate guarding in males (Alberts *et al.,* 1996; Swedell *et al.,* 2008) and costs of gestation and lactation in females can lead to changes in their foraging activity (Muruthi *et al.,* 1991). Pregnant or lactating females will spend more time feeding than cycling females (Muruthi *et al.,* 1991). Staying vigilant of predators and transporting infants can affect a females' ability to spend time foraging (Altmann, 1980; Whiten, 1982; Muruthi *et al.,* 1991; Alberts *et al.,* 1996). Even without factoring in reproductive costs, individual male and female baboons may still forage differently due to body size related nutritional needs.

1.2 Problem Statement

The ability to find and exploit food aggregations by baboons and the varying nutritional needs between adult male and female baboons can affect the amount of Opuntia eaten by Namu (NMU), the study troop. At the time of this study, Opuntia distribution was spatially restricted throughout NMU's home range, meaning that some trade-off relationships may exist between Opuntia fruits and other important food resources. These factors were evaluated to determine what impacted the troops' Opuntia usage by conducting real time focal follows for eight adult baboons in the troop.

1.3 Justification

The Uaso Ngiro Baboon Project (UNBP) has documented the spread of Opuntia into the study troop's home range (Strum et al, in press), and has collected data on fruit handling and preparation techniques by individual baboons. The manipulation required for the fruits that are ingested is costly, in that it takes time and energy to prepare. Opuntia patches are in specific areas of the range, originating near Dol Dol, away from the study troop's home range. Their home range is at the front of the invasion where the Opuntia patches are smaller, fewer, and unevenly distributed. It takes time and energy for the troop to reach these distributions. This study was a unique opportunity to investigate some factors that influence how a relatively new and energy-rich food source contributes to the diets of these olive baboons.

1.4 Objectives

1.4.1 Main Objective

The purpose of this study was to investigate how Opuntia fruits contribute to the foraging strategy of individuals from a single troop of baboons in Mukogodo Division, Kenya.

1.4.2 Specific Objective

1) To determine how daily foraging strategies of olive baboons relates to variations in Opuntia fruit selection in Mukogodo, and

2) To assess variation in diet between adult male and female olive baboons.

1.5 Hypotheses:

H_1:   Baboons will typically eat the same amount of Opuntia fruits during different times of the day.

H_2:   Adult male and adult female baboons within the same age class will not show any variation in the amount or type of Opuntia fruits eaten, regardless of their varying nutritional needs.

CHAPTER TWO

Literature Review

2.1 Foraging behaviors of Savannah Baboons

Baboons are the most widespread of the African primates, found in arid grasslands and savannah, as well as forests and mountain regions (Estes, 1991; Groves, 2001). Olive baboons (*Papio anubis*) are spread throughout equatorial Africa, distributed over 25 countries. Their range extends south from the Sahara into Kenya, Democratic Republic of Congo, Tanzania, Uganda, Burundi, and Rwanda (Estes, 1991; Cawthon, 2006). Baboons are opportunistic omnivores and have a highly flexible diet, allowing them to exploit a variety of habitats. They utilize virtually all plant species available within their range, including grasses, tubers, blossoms, seeds, seed pods, insects, shoots, fruits, nuts, and flowers (Altmann & Altmann, 1970; Post, 1982). During dry seasons they are known to supplement their diets with small vertebrates such as Cape hare, vervet monkeys, gazelle fawns, reptiles, beetles, termites, birds, and bird eggs (Strum, 1975; Post, 1982).

2.2 Optimal Foraging Theory (OFT)

Foraging behavior includes the mechanisms and decisions animals use to maximize energy gain and how an animal searches for, locates, processes, and consumes food resources from the environment (Cant and Temerin, 1984; Mellgren and Brown, 1987; Walker *et al.,* 1999; Segal, 2008). Foragers typically make decisions based on spatial distribution of resources, nutritional content, rates of resource renewal, and seasonality (Garber, 1987; Wahungu, 1998; Jaman *et al*, 2010). OFT states that in cases of food scarcity the forager will either i) specialize on a single food type, if that food is in abundance, when the energy intake is greater than expenditure (energy maximizing) (Charnov, 1976; McFarland, 1994; Mutua, 2001); or ii) the forager will stay in a patch as long as that patch yields the greatest intake (time minimizing) (MacArthur and Pianka, 1966; Garber, 1987; McFarland, 1994; Mutua, 2001). Primates have been known to exploit food patches or aggregations instead of an individual food source (Barton, 1989; Mutua, 2001; Sayers *et al.*, 2009), but can typically alter their foraging behavior depending on nutritional requirements and food availability as they encounter resources randomly and rank them according to their potential energy contribution (Sussman, 2000; Boyer *et al*, 2006; Jaman *et al*, 2010; Bridgeman, 2012). Baboon species are characterized by a large degree of variation in dietary selection and foraging behavior linked to ecological factors, climatic conditions, and nutritional value (Hill and Dunbar, 2002; Okecha and Newton-Fisher, 2006; Jaman *et al*, 2010), as well as temporal and spatial patterns of available food (Glander and Teaford, 1995; Wahungu, 1998; Janson and Chapman, 1999; Gemmill and Gould, 2008), and topographical conditions (Mutua, 2001).

A study conducted by Hill and Dunbar (2002) showed that baboon diets across many habitats in Africa were composed of mostly fruits, leaves, subterranean items, and to a lesser extent, flowers and animal matter. For baboons sampled in Chololo, Kenya, an area adjacent to the baboons sampled in this study, the majority of food consumed consisted of fruits (23%) and leaves (27%), followed by flowers (21%) and subterranean items (15%) (Barton, 1989; Hill and Dunbar, 2002). Recent studies (unpublished data, UNBP) have shown a large part of their diet is now being supplemented with the fruits and seeds of *Opuntia stricta*, a succulent plant species commonly known as the 'prickly pear cactus'. Opuntia is an invasive species that has been widely dispersed by elephants, baboons, and humans, and is concentrated in dense patches in some areas throughout the Mukogodo range. Opuntia bears fruits throughout the year and is therefore abundant even in the dry season, or in times of food scarcity. Opuntia fruits require some manipulation before consumption to remove the small hair-like bristles (*glochids*) on the peel, and the peel itself. Seeds must be harvested indirectly out of elephant dung (unpublished UNBP data). Adult baboons can reach fruits from the ground unless the plant is particularly large. Peels and dry fruits can be picked off the ground easily by baboons of all age/sex classes. Opuntia fruit harvesting does require some energy expenditure but baboons can readily eat Opuntia fruits when they are available.

2.3 Hierarchal Constraints to Diet

Multi-male baboon societies have complex hierarchies that have separate rules for adult males and adult female members of the troop. Daughters of female members will usually occupy the rank immediately under their mother, with juvenile daughters of high-ranking females dominant to adult females of a lower rank. Therefore, female hierarchies are usually stable over many generations (Bercovitch and Strum, 1993). Resource holding potential refers to an individuals' ability to gain and maintain control of a particular resource (Silk, 2002). Resource holding potential extends to diet, when preferred foods are considered an important resource. Matrilineal lineage is just one factor in an individuals' resource holding potential. Males exist within a more fluid hierarchy than females. Males rely more heavily on fighting over resources due to an uncertainty about their resource holding potential (Silk, 2002). This uncertainty is caused by both their ever-changing hierarchy and because they are the dispersing sex (Silk, 2002). In multi-male troops the males will fight over access to females, as well as preferred foods and other resources. Besides acts of aggression, a lower ranking member of the troop can simply avoid the dominant member, thus avoiding conflict. This is true for both males and females, as females will also engage in aggressive behavior. An individual of low rank will therefore have more constraints on his or her diet than a high-ranking individual.

2.4 *Opuntia stricta* in the Rangelands

The following information on *Opuntia stricta* was adopted from BioNET fact sheets and the Invasive Species Compendium database (CABI, 2011), unless otherwise referenced:

O. stricta, commonly known as the Prickly Pear cactus, is invasive in many parts of Kenya (Central Kenya, Tsavo East) and is present in Uganda and Tanzania. It was first introduced to the African continent in the 1950's. The continuing deterioration of the rangelands in Mukogodo has allowed the plant to spread easily. It grows upright to about 2 meters tall, with many-branched stems consisting of fleshy, succulent pads and segments (*cladodes*). Flowers are a bright yellow when in bloom. Immature fruits are green, turning a greenish purple color during transition. The fruits are ripe when they are deep red to a purple color. The fruits are covered in small, hair-like bristles (*glochids*), which must be removed before eating. Seed to fruit abundance is high, which makes it easily dispersible (Natural Resources Conservation Service, US Department of Agriculture, 2011). In Mukogodo, seeds are primarily distributed by baboons, elephants, and livestock, which are spread through their droppings. Humans use Opuntia as fences and children eat the fruits which contribute to seed distribution.

2.5 Opuntia Nutrition Data

Previous studies of diet selection in baboons have shown that nutritional content plays a large part in the selection process. Protein, fiber, phenolics, alkaloids (Whiten *et al.,* 1991; Barton and Whiten, 1994), water (Barton, 1989), micronutrients (Gaynor, 1994), and carbohydrate levels (Altmann, 1998) have all been cited as significant factors (Hill and Dunbar, 2002). Opuntia fruits have very high moisture content at about 90%, carbohydrate levels are about 8%, and each fruit contains 35Kcal/g of energy. Much of the fiber is contained in the peel (7.7%) and the majority of the protein and phenolic content lies in the seeds (2.1% and 16.5 mg/g, respectively). The peel contains a slightly higher content of Vitamin C (60mg/g) although the pulp does contain a substantial amount at 57mg/g (Kunyanga and Imungi, 2009). The pulp holds the majority of the water content. The majority of fiber is in the peel, which is usually discarded by baboons during consumption. Ripe fruits are easily metabolized and absorbed into the large intestine (Segal, 2008). They are rich in carbohydrates and sugars, which make them a good source of energy (Garber, 1987). Ripe fruits and seeds are both generally high in carbohydrates and protein (Altmann and Alberts, 1987; Barton *et al.,* 1992; Gaynor, 1994; Heller *et al.,* 2002; Kunz and Linemair, 2007; Segal, 2008), however the seeds cannot be digested by baboons nor can they be directly harvested. Opuntia seeds have to be harvested out of elephant dung for the baboons to gain any protein from them (unpublished UNBP data). An individual selecting food on the basis of nutritional content may select Opuntia fruits depending on which nutrients are needed at that time (energy versus protein or fiber). In this case, the time of day may factor in as to when individuals eat Opuntia.

CHAPTER THREE

Materials and Methods

3.1 Study Area

3.1.1 Location

The study area (Figure 3.1) is located on the Laikipia plateau in central Kenya, East Africa, about 250 km north of Nairobi and 84 km north of Nanyuki town, measuring approximately 50 km^2 (Mutua, 2001). The estimated home range for NMU is also shown in Figure 3.1. Climatically, the area falls into the category of "dry savannah" (Delaney and Happold, 1979). Mean annual rainfall is between 350 mm and 450 mm (Strum, 2005). Temperatures during the months of August 2011 and February 2012 were measured to be between 9°C and 36°C, with higher temperatures recorded in the dry season (UNBP rainfall data, 2011-2012). The area experiences a bimodal rainfall distribution consisting of two rainy seasons; the shortest period in November and a longer period between March–July (Mutua, 2001.)

Between August 2011 and February 2012 the monthly mean biomass for Chololo Ranch fluctuated greatly, ranging from 20 g/m^2 to almost 100 g/m^2 (Figure 3.1). The biggest spike in biomass was during December and January. This caused there to be some residual herb layer in February, about 40 g/m^2, which is unusual for this area.

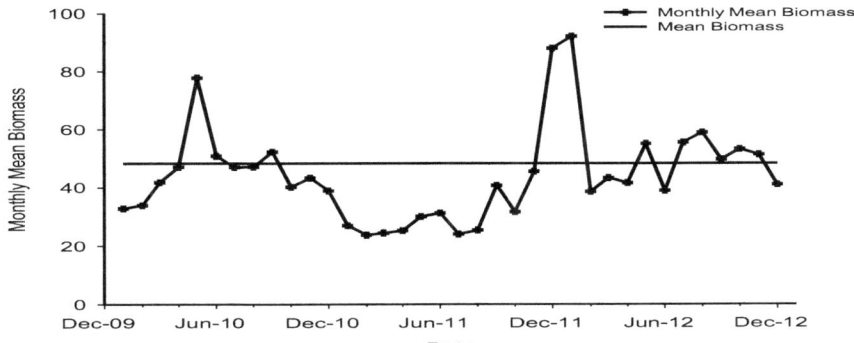

Figure 3.1. **Monthly mean biomass for Chololo Ranch. The period between August 2011 and February 2012 was extracted for this study. (UNBP ecological monitoring data, 2012).**

3.1.2 Soils and Vegetation

The soils are moderately deep to deep clay loam, medium to high water retention capacity, and have a high natural fertility (Mutua, 2001). Vegetation consists of *Acacia etbaica, A. tortilis, A. nilotica* and *A. mellifera* woodlands, grasslands made up of mostly *Cynodon, Tragus bertorianus,* and *Penissetum,* with *Kyllinga* species in some areas especially after the rains (Barton and Whiten, 1994; Mutua, 2001). Succulents such as *Sanseveria* and several *Opuntia* species also dominate much of the home range (Barton and Whiten, 1994). Several high rocky outcrops or 'kopjes' are used by the baboons for sleeping sites (Barton and Whiten, 1994; Mutua, 2001).

3.1.3 Fauna

This area supports a wide variety of large mammals, primates, and carnivores. Large mammals such as elephants, buffalo, reticulated giraffe, zebra, Thompson gazelle, and

Grant's gazelle are present (Barton, 1989; Mutua, 2001). Other primate species include vervet monkeys and lesser bushbabies (Barton 1989; Mutua, 2001). Predators include leopards, hyenas, lions (Strum, 2005), and wild dogs. Humans are also a threat and contribute to baboon mortality (Strum, 2005).

3.1.4 Land Use and Conservation Status

Mukogodo Division, within Laikipia County, supports both livestock and agriculture production and consists of both community owned lands and private ranches (Laikipia Wildlife Forum, 2013). In Mukogodo economic livelihoods include activities such as sand harvesting, charcoal production, and livestock production. Mukogodo was designated as a National Forest Reserve in 1937 (Survey Branch of Kenya Forest Department, 2012). There are other sanctuaries in the area, such as Ol Jogi and Chololo Ranch, which serve as designated areas of research.

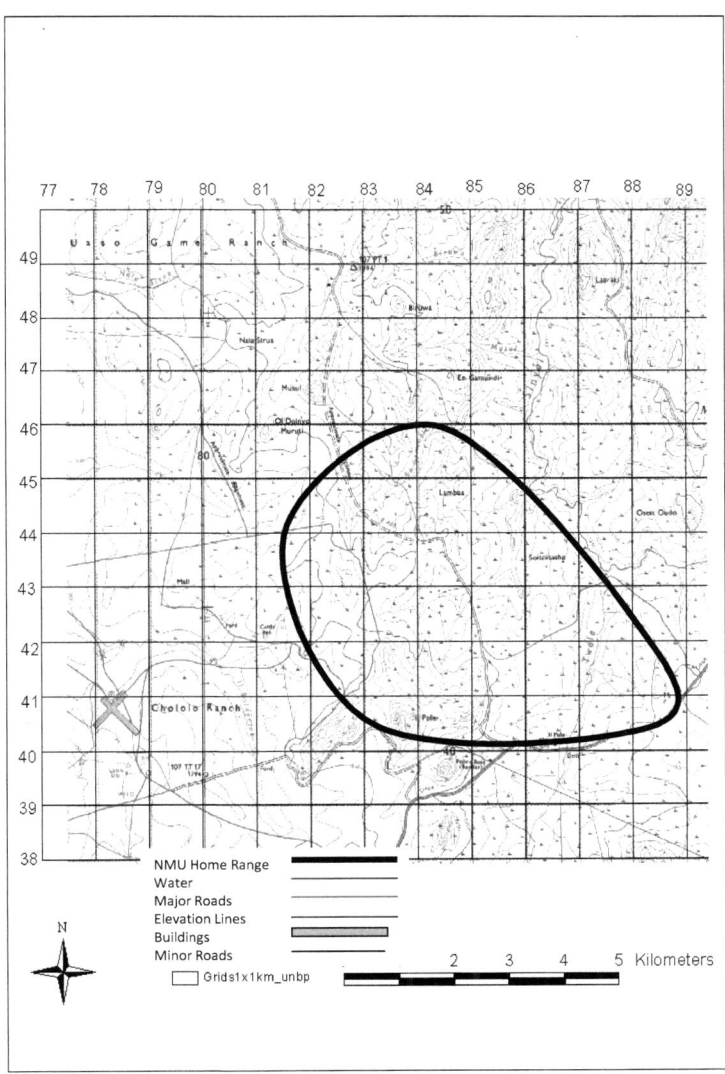

Figure 3.2: Map of the study troop's home range in Mukogodo Division, Kenya. The estimated home range for NMU is outlined in black.

3.2 Study Baboons in Namu Troop

In 1984, three troops of wild baboons were translocated from Kekopey ranch in Gilgil, Kenya to the Laikipia plateau, a distance of about 200 km to the north from the original location (Strum, 2005). The translocated group consisted of Pumphouse Gang (PHG), Malaika (MLK), a daughter troop of PHG that had fissioned in 1981, and Cripple (CRIP). Some original members of MLK now make up part of NMU, after fusing with Soitoitashe (STT), an indigenous troop. NMU is a daughter troop of MLK and STT, declared fused in September 2001 (UNBP records). The translocation was deemed successful based on the birth and mortality rates after translocation (Strum, 2005).

A total of eight individual baboons (four males and four females) were selected from NMU for this study. The troop fluctuated between 77 – 84 members over the study period. The approach of this study was to look at males and females as singular "case studies" as opposed to a generalized statistical sample. I therefore conducted longer day follows instead of frequent short samples. Members of the troop were identified using external markings and physiognomy. The four males (H1, NP, PU, XV) were selected out of ten others in the same age-class, classified as sub-adult 3. This means that the males had almost reached their full size (aged between 8-10 years) (Strum, 1982; Bercovitch, 1989; Alberts and Altmann, 1995), had enlarged testes signaling sperm production (Alberts and Altmann, 1994), and were therefore considered able to father offspring. This classification is based on UNBP protocol definitions and records. It was assumed that their nutritional needs would not vary between them, since they are all in the same age class and reached roughly the same point in their growth. The four females (KT, SB, TL, TO), selected out of 23 adult females, were chosen on the basis of being in the same reproductive state. They were sexually cycling and did not have dependent infants at the start of the study. All four females became pregnant during the study, and were reclassified as such in October 2011 (UNBP records). Because they became pregnant at roughly the same time their nutritional needs would change similarly, so no alterations were made to the selected female roster. The troop was already well-habituated, having been frequently used as a study troop by UNBP since 1984.

NMU's home range is north of the village of Il Polei, northeast of Chololo Ranch, triangulated between three main sleeping sites: White Rocks (WTR) to the north, Sisal (SIS) to the south, and Bridal (BRD) in the east. The majority of Opuntia is located around BRD and to the east of SIS, while herb layer and tree foods are found around WTR. The banks of Twala Gully (TWG), a dry river, also produce a dense layer of grasses and succulents. Herb layer emerges after rains, consisting of *Ammocharis* herbs, *Penissetum*, and *Cynodon* grass species. Herbs like *Tragus terrestrius* and *Oxygonum sinuatum*, and sedges like *Kyllinga allata* and *K. alba* are more available closer to WTR. *Sanseveria abyssinica, S. intermedia* and *Acacia* species are found throughout, with *Acacia xanthophloea* located only in the gullies of dry rivers.

3.3 Sampling Design

This study was conducted between August 2011 and February 2012. Individual baboons were followed based on a rotating schedule, selected randomly in the morning at the start of each focal follow. No individual baboon was repeated until all eight had been followed, constituting a single rotation. The location of the troop was determined every night based on their sleeping site. A "morning" focal follow lasted between 5- 6 hours from when the baboon descended a sleeping site and started foraging. "Full day" focal follows lasted between 9-10 hours starting from when the baboon descended a sleeping site, began foraging, and then ascended a sleeping site in the evening (morning follows = 53; full day follows =10; total sample size = 63). Focal follows that did not last at least five hours for morning follows or nine hours for the full day follows were omitted. This was done to maintain homogeneity. Full day follows were conducted roughly about once a week in order to assess evening foraging behaviors, and each individual baboon was observed at least once for an entire day. Full day follows were not collected as frequently since it was often difficult to stay in the field during the rainy months of October and November.

Any individual involved in a consort was not followed until the consort had finished. Consorting baboons deviate greatly from their regular feeding patterns, often ranging very far from the main part of the troop. Consorts were well-documented to keep track of which males and females were involved, especially those individuals selected for this study. This sometimes affected when individuals were followed. However there were enough animals selected for the study such that no individual was chosen more frequently than any other. If the selected individual for that day was found to be in consort, another baboon whom had not yet been followed in that rotation was chosen at random (See Appendix 1 for the sampling schedule).

3.4 Data Collection Techniques

Individuals were observed using real-time focal follows, as opposed to shorter focal or scan follows (Okecha and Newton-Fisher, 2006), to construct diets. A focal follow consists of choosing one individual baboon and staying with that individual for the entire observation period. This gives a more accurate representation of the baboon's activity and diet over the course of a day, producing a longer and more representative picture. Baboon studies typically use scan sampling methods as outlined by Altmann (1974). For this study real-time focal follows were used to represent individual "case studies" instead of shorter scan follows which would require a greater number of samples to produce a "statistical picture" (see for example in chimpanzees, Watts and Mitani, 2000; Corp and Byrne, 2001;

Basabose, 2002). Data was recorded on a check sheet to measure the amount of time spent on different food types. All species of plants that were eaten were recorded using codes from the UNBP Food List (see Appendix 2), which included the four Opuntia species found in the home range (*stricta, subulata, vulgaris,* and *ficus-indica*). The number of *O. stricta* fruits eaten was counted individually; all other food types were measured in seconds and minutes. Follows were measured in five-minute intervals, recording the individuals' activity at the start of each interval. Activity codes were used in conjunction with existing UNBP codes (see Appendix 3).

Classification of food types were grouped as follows: grass/herbs, bush/trees (bush or tree leaves and flowers), *Acacia* tree foods (*Acacia* seeds and seed pods), miscellaneous foods (to include other succulents like *Sanseveria* in the area, invertebrates, and unidentified food items), and Opuntia fruits. Time was measured by distinguishing between 'instantaneous feeding' and 'feeding bouts' on one specific food type. 'Instantaneous feeding' constituted an individual picking a specific food type or species one time only, and marked as an X on the data sheet. 'Feeding bouts' described individuals spending more than one minute on the same food type or species. Full minutes were marked by tallies as opposed to X's for clarification. Individual bites or hand-fulls were not counted; instead the duration of time spent feeding was measured (see Appendix 4 for total feeding times for plant species eaten each month). By contrast, Opuntia fruits eaten were counted individually rather than marking time spent on each fruit. Ripeness and size were also noted for each Opuntia fruit picked using current UNBP protocol (see UNBP Food List in Appendix 2).

3.4.1 Foraging selection frequency

Dietary composition was constructed from the data for each month and tested for significant differences in food choices through the study period. All focal follows were examined in one hour blocks to find the percentage of time spent feeding per hour, and percentages of time spent feeding on each food type per follow to determine if certain foods were preferred at different times of the day. The percentages of each hour and the percentages of each food type calculated per follow were used in analysis. A rate of Opuntia fruits eaten per minute was calculated from the total fruits consumed per hour in order to compare directly with the percentage of time spent on the other food types. The total number of fruits eaten was divided by the number of hours in each follow, and included in the total time spent feeding. This rate was used to compare directly with the time spent on other food types and defined as 'Opuntia usage'.

The variation in foraging between the followed males and females was examined by taking the monthly averages of time spent on each food type, and the number of fruits consumed. The averages were compared against each individual and by sex.

3.5 Data Analysis

All data sets were first tested for homogeneity and normal distributions. When these conditions were satisfied, data was analyzed using ANOVA or paired t-tests. Non-parametric Spearman's rank correlation and Kruskal-Wallis tests were done for any heterogeneous data. All analyses were conducted at a 0.05% significance level.

All data measuring time spent on food types were standardized using percentages from the totals in each follow. Percentages of time spent on food types, combined with the rate of Opuntia eaten per minute, was used in the analysis to control for error due to the uneven amount of follows between morning and full day sets. For each follow, the total time spent on each food type was divided by the total time spent feeding. The hourly breakdown for each follow was done to test for variability within one follow or variation in one day. The minutes spent on each food type and the Opuntia fruits eaten per minute were added and the total divided by the number of minutes observed to get a percentage of total time spent feeding. The variables for 'time of day' were set as three values: 'morning', 'mid-morning', and 'evening'. Morning and mid-morning values covered three hours of data each ('morning'= first three hours of each follow; 'mid-morning' = subsequent three hours), while the evening value covered the last four hours of the full day follows. The morning follows are therefore only comprised of 'morning' and 'mid-morning' values. The levels of Opuntia fruit ripeness were classified as ripe, unripe, and dry. Unripe fruits included all green and transitional fruits.

ANOVA was used to test for significant differences on the time spent on the different food types between months. Hourly percentages and the number of whole Opuntia fruits eaten per hour were compared and analyzed separately using ANOVA to examine daily diet patterns. One-way ANOVA was used to analyze monthly variation in diet between adult males and female baboons.

Paired t-tests were used to test mean differences between males and females for the following: in the percentage of total time spent feeding on all food types each month; variation in the mean number of Opuntia fruits eaten, the mean time spent on grass and herb species, and mean time spent on bush and tree species each month.

Spearman's rank test was used to analyze the relationships between the percentages of the total time spent on each food type against the rate of Opuntia eaten per minute, and for correlations between the number of fruits eaten per hour and the minutes spent feeding per hour each month.

Kruskal-Wallis ranking tests were performed for monthly variations in time spent feeding per hour for all sampled individual baboons.

CHAPTER FOUR

Results

4.1 The role of Opuntia in the foraging behavior for olive baboons in Mukogodo

4.1.1 Dietary Composition

The dietary composition for the study troop for the entire study period from August 2011 to February 2012 is shown in Figure 4.1. Grass and herb species were eaten the most overall (52% S.E.±1.87), followed by Opuntia fruits eaten per hour (25% ±0.94), bush and tree species (12% ±4.82), and miscellaneous and other succulent species (11% ±1.75) (labeled as 'all other foods'). Seeds and seed pods had the lowest mean percentage (1% ±0.00). The mean percentage of time spent on individual food types was significantly different (ANOVA, $F_{4,285} = 70.43$, $p < 0.0001$).

Figure 4.2 shows that grass and herb species were eaten more than any other food type each month. The rate of Opuntia fruits eaten per minute was fairly even throughout, decreasing to a mean±S.E. of 19±0.02% in January and 16±0.01% in February. Figure 4.2 also illustrates the inverse relationship between food types, particularly between grass and herb species and Opuntia fruits eaten per minute. Significant correlations existed between several food types. The rate of Opuntia fruits eaten per minute was inversely correlated with grass and herb species (Spearman's rho, $r = -0.616$, $n = 63$, $p < 0.0001$), bush and tree species (r = -0.299, n = 63, p = 0.017), and positively correlated with succulents and miscellaneous species ('All other foods') (r = 0.314, n = 63, p = 0.012). Significant inverse relationships were also found between grass/herb species and succulent/miscellaneous species ('All other foods') (r = -0.562, n = 63, p < 0.0001), and bush/ tree species (r = -0.271, n = 63, p = 0.032).

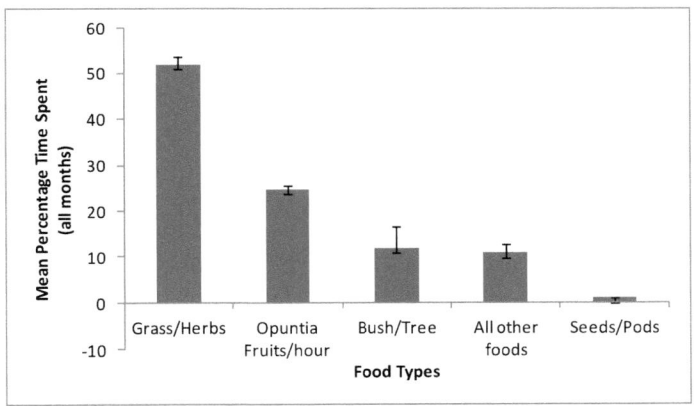

Figure 4.1: The mean percentages of time spent on each food type from August 2011 – February 2012.

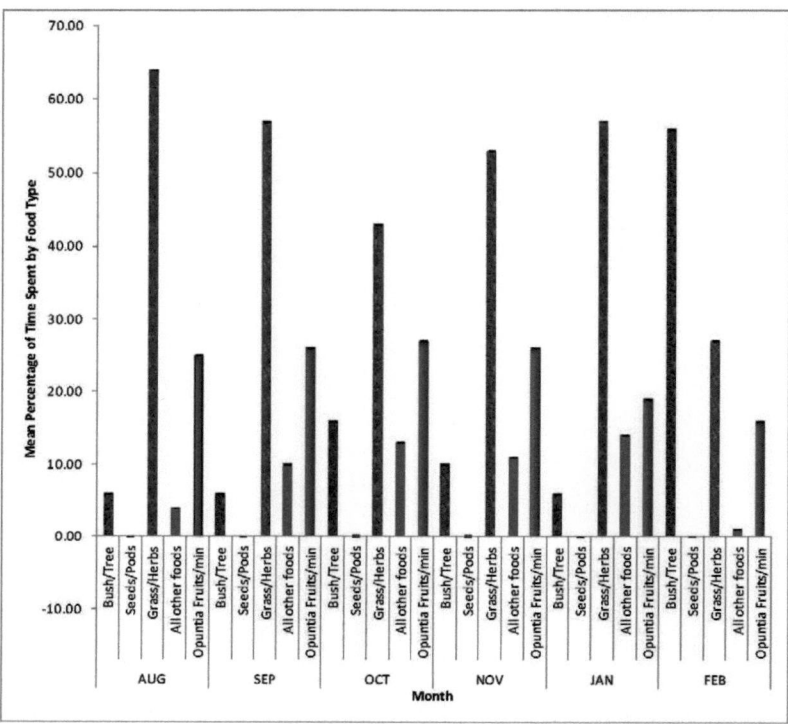

Figure 4.2: Monthly mean percentage of time spent on different food types by the study troop.

4.1.2 Daily Diet Patterns for Opuntia fruits

The number of whole Opuntia fruits consumed per hour was analyzed for differences between months and between the times of the day. Figure 4.3a shows that between months there was not much difference between the mean numbers of fruits eaten. From August to October, there was a slightly increasing trend; the maximum mean fruits eaten per hour were in October (17.44 ± 3.97 fruits/hr). Over the course of a day, the baboons did not have a preference for which time of day they consumed the fruits (Figure 4.3b). ANOVA analysis confirmed there were no significant differences for either variation between months or time of day (month, $F_{5,400} = 1.722$, $p = 0.128$; time of day, $F_{2,400} = 0.068$, $p = 0.934$).

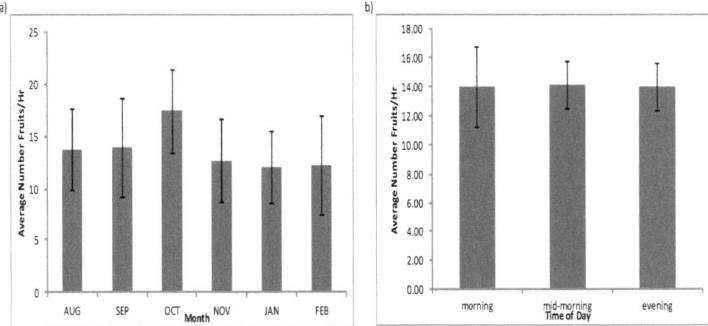

Figure 4.3: Average number of Opuntia fruits eaten per follow (fruits/hour) by the troop (a) from August 2011 through February 2012, and (b) over the course of a day.

4.2 Variation in nutritional requirements or time-energy budgets for adult male and adult female baboons

4.2.1 Food type selection and feeding time

There was variation in the mean time spent on each food type between sexes. Figure 4.4 shows that females spent on average more time on each of the food types than males each month. January is the only exception; where males spent more time on grass and herbs, and bush and tree species. Females spent a greater percentage of time feeding but not to a significant level ($t = 1.367$, df $= 24$, $p = 0.184$). Overall, females spent significantly more time on grass and herb species (Paired t-test, $t = 2.836$, df $= 25$, $p = 0.009$) than males.

ANOVA showed that the time spent on different food types were significantly different for males ($F_{4,25} = 10.419$, $p < 0.0001$) and females ($F_{4,25} = 5.833$, $p = 0.002$) but this was not significantly different from month to month ($F_{5,62}$ = 1.748, p = 0.138). Multiple comparison tests indicated that both males and females spent the most time on grass and herb species, than all the other food types as shown in Table 4.1.

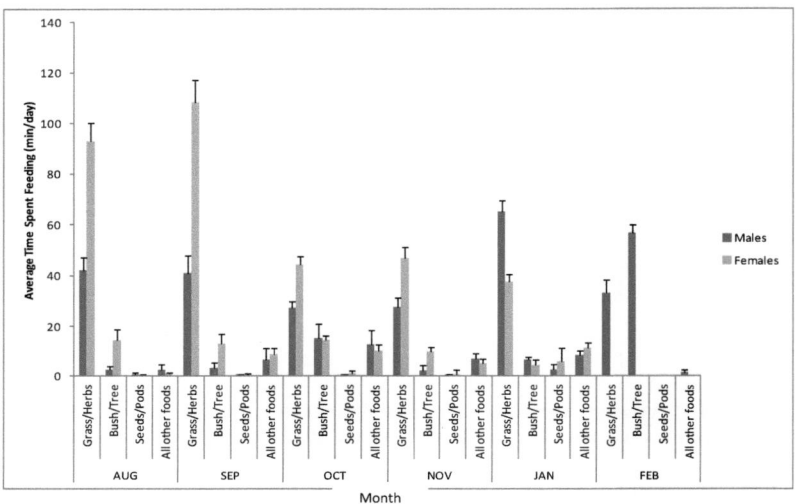

Figure 4.4: The average minutes spent per day on different food types, excluding Opuntia fruits, eaten during August 2011 through February 2012 for both males and females. There were no females followed in February as all four had already been followed.

Table 4.1: Homogenous subsets for the mean time (minutes) spent on each food type from August 2011 through February 2012 for male and female baboons.

Males (Minutes Feeding)					Females (Minutes Feeding)			
Food	N				Food	N		
		1	2				1	2
seeds_pods	6	2.30			seeds_pods	6	7.90	
all_others	6	39.08			all_others	6	42.52	
bush_tree	6	60.57			opuntia_fruits	6	56.02	
opuntia_fruits	6	95.61			bush_tree	6	56.60	
grass_herb	6		203.08		grass_herb	6		332.68

4.2.2 Opuntia fruit selection

Males ate significantly more individual Opuntia fruits than females each month, as shown in Figure 4.5 (Paired t-test, t = 5.383, df = 25, p < 0.0001). ANOVA showed significant differences in mean percentages of fruits chosen based on ripeness for males and females. Multiple comparison results are shown in Table 4.2. Both males and females preferred riper fruits. Males chose ripe fruits significantly more than unripe and dry fruits. Females chose ripe fruits and unripe fruits significantly more than dry fruits (males, $F_{2,38}$ = 290.348, p < 0.0001; females, $F_{2,38}$ = 7.718, p = 0.002).

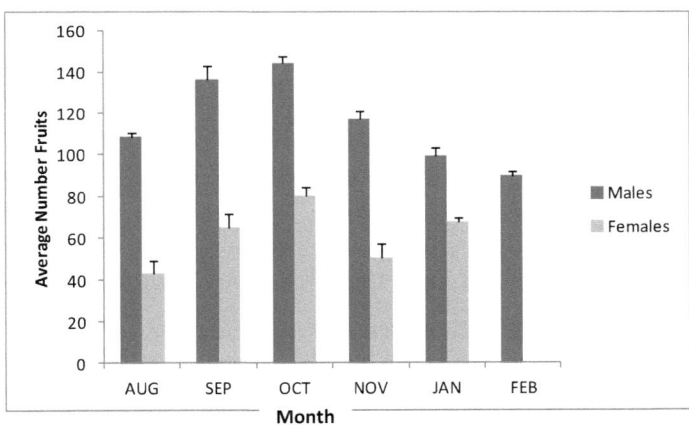

Figure 4.5: Average number of Opuntia fruits eaten by baboons from August 2011 to February 2012. No females were followed in February as all four had already been followed.

Table 4.2: Percentages of fruits chosen based on ripeness per follow between males and females from August 2011 through February 2012.

Fruit Preference for Males					Fruit Preference for Females			
ripeness	N	Subset			ripeness	N	Subset	
		1	2				1	2
dry	3	0.030			dry	3	0.0233	
unripe	10	0.095			unripe	18		0.5640
ripe	28		0.9407		ripe	30		0.7832

4.3.4 Impact on Opuntia fruit use and feeding times

Comparisons of feeding time by month did have some variation (Figure 4.6a). There was a negative parabolic trend; with the troop spending more time feeding per hour in August and September (mean time ± S.E. = 12.09±3.60 min/hr, 12.36±3.42 min/hr, respectively), decreasing in November (mean time = 8.00±2.53 min/hr), then increasing again in January and February (mean time = 10.83±3.04 min/hr, 12.36±3.19 min/hr, respectively). Baboons spent the greatest amount of time feeding during the evening hours (Figure 4.6b), significantly more than the other blocks of time (Kruskal-Wallis, $H = 8.730$, n= 417, $p = 0.013$). Multiple comparison tests confirmed that the time spent feeding in the evening was significantly higher than time spent feeding in the morning hours, with the mean time spent feeding increasing as the day progressed. There was no significant monthly difference in time spent feeding per hour (Kruskal-Wallis, $H = 7.364$, n = 417, $p = 0.195$).

A correlation analysis was done to test for associations between the number of Opuntia fruits eaten per hour and the minutes spent feeding per month. An inverse relationship exists, but not significantly (Spearman's rho, $r = -0.093$, n = 417, $p = 0.057$).

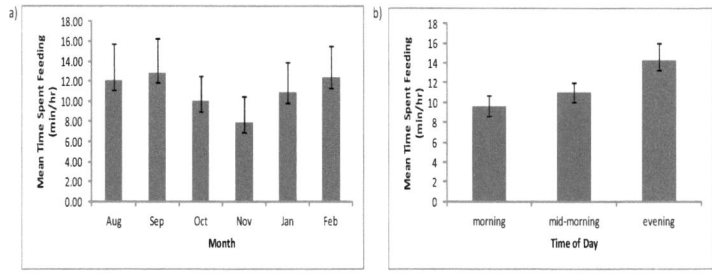

Figure 4.6: Mean time spent feeding per follow (min/hour) by the troop (a) from August 2011 through February 2012 and (b) over the course of a day.

CHAPTER FIVE

Discussion

5.1 The contribution of Opuntia fruits in the daily foraging behavior for olive baboons in Mukogodo.

Opuntia fruits contributed to the baboons' diet by the second largest percentage after grasses and herbs. Analysis showed a significant trade-off relationship between grass and herb species and Opuntia fruits eaten per minute. During months where the percentage of grass and herb species decreased, the total Opuntia fruits eaten per minute increased, and vice versa. The baboons did not focus on Opuntia fruits at any particular time of the day, but did eat more fruits in the month of October. Analysis showed a negative parabolic trend in overall feeding time per hour over the course of the study. Evening hours had the highest mean time feeding per hour, meaning that the troop spent the hours before ascending a sleeping site feeding and foraging. During October and November, as rain began to fall, the mean time spent on all the food types was more evenly distributed.

The troop appeared to employ a mixed foraging strategy. Some months the troop was observed spending more time in one area and utilizing a specific food resource for as long as possible, thereby minimizing their time spent doing other activities such as traveling. This behavior was observed when there was an abundance of herbs and grasses, mainly during January when there was a spike in biomass. In February when the herb layer became dry the troop still managed to utilize the herb layer by eating other parts of the same grass and herb species (i.e., corms, stem bases), and eating more Opuntia fruits to supplement moisture. Because the main aggregation of Opuntia is on the far eastern side of NMU's home range, the troop had to travel some distance to incorporate the fruits. This suggests that during months when herb layer was not eaten as much, the energy spent traveling was made up for by the extra calories and moisture found in the Opuntia fruits, hence maximizing their total energy intake. The trade-off between Opuntia fruits and herb layer, the two food types with the highest percentage of utilization, suggests that Opuntia fruits are an important food resource for the baboons in NMU. The extra caloric and moisture intake from the fruits (Kunyanga and Imungi, 2009) allow the baboons to almost solely focus on grass and herb species and Opuntia fruits. The other food types in the area seemed to be utilized as a supplement to their daily and monthly diets.

5.2 Variation in nutritional requirements or time-energy budgets for adult male and adult female baboons.

Significant differences between sex and food types were found, as well as differences among sex. Both sexes preferred grass and herbs. Males chose Opuntia fruits second after grass and herbs, while Opuntia fruits ranked third for females. The difference in food preference could be based on nutritional requirements due to body size and reproduction costs. In October, halfway through the study, all the female baboons involved in the study became pregnant. The pregnant females ate more herb layer than males possibly to ingest more protein to deal with the costs of gestation (Strum, 1991; Alberts *et al.,* 1996; Swedell *et al.,* 2008). Males ate more Opuntia fruits probably to satisfy their higher caloric needs, as past studies would suggest (Altmann and Alberts, 1987; Strum, 1991), in accordance with their larger absolute body size. For example, adult males are twice the size of adult female baboons and therefore have higher metabolic needs (Muruthi *et al*, 1991). It is difficult to say from this study alone if the females' pregnancies truly influenced their diet selection with respect to their male counterparts since there were no sexually cycling females to compare them to.

Significant variations existed for preferences of fruits based on ripeness between and among sexes. Males had a higher rate of selection for riper fruits, whereas the females had a higher rate of selection for unripe fruits. The females could then be choosing more unripe fruits based on availability, since it is known that unripe fruits are harder to digest, aren't nearly as palatable (Segal, 2008), and have less nutritional value (Garber, 1987). The placement of the fruits on the plants could therefore be a factor. Ripe fruits tend to be at the top of the Opuntia plant and would have been out of reach for the females due to their smaller body size. A passing adult female could take what she could in order to get what calories she could from the less desirable but more available Opuntia fruits. A foraging male, on the other hand, would be able to take his time and chose a ripe fruit higher up the plant. The difference in body size ratios between adult males and females again becomes a factor in baboon foraging behavior.

Comparing males and females by month it became clear that the females spent more time feeding per follow. The total times feeding in August and September served as a control to see if the increase was caused by the pregnancies and their increased nutritional needs. Even during those months the females spent a greater amount of time foraging. The difference in male and female foraging percentages in August and September serve as a baseline comparison, illustrating that these females may still generally feed more than the males under normal circumstances. It was observed during the follows that the females would spend more time harvesting Opuntia fruits. They were much pickier about cleaning each individual fruits, rolling each fruit on the ground and wiping them clean before gently peeling or squeezing the fruits out of their skins. This prolonged harvesting technique would often result in less fruit consumed in one sitting. The benefits of prolonged cleaning meant that the fruit would be easier to digest, an important factor for females since they have smaller, more sensitive digestive tracts (Altmann and Alberts, 1987; Strum, 1991; Strum, 2009). The males were observed sometimes taking whole fruits two-by-two with little notice of the *glochids*, tear the top off and consume the entire fruit, dropping the empty skin out of their mouths when finished. This behavior was observed usually after rains when there were less *glochids* on the fruits. Variations in harvesting technique were an observable difference only.

CHAPTER SIX

Conclusions and Recommendations

6.1 Conclusion

The baboons in Mukogodo Division employed a mixed foraging strategy; spending more time travelling between the areas with high densities of herb layer and Opuntia distribution some months, or saving their energy and specializing on either herb layer or Opuntia fruits during other months. During drier months, the troop focused their time and energy on predictable foods and aggregations, spending greater amounts of time in the areas with either herb layer or Opuntia plants, and therefore spent more time feeding. This trade-off relationship is inferred since true seasonality was not observed during the study period. The troop spent the evening hours before ascending a sleeping site foraging more than any other time of the day.

Males and females utilized different foods at different rates due to their differing energy needs. To cope with their increased gestational energy requirements, all four females foraged for longer intervals than their male counterparts. Females ate more grass and herb species, as well as bush and tree foods, to ingest more protein; males ate more Opuntia fruits to satisfy their greater caloric needs. Differences in selection of Opuntia fruits based on ripeness also existed; males chose more ripe fruits, while females ate a higher percentage of unripe fruits. Variation in fruit selection by females was probably influenced more by overall fruit availability and accessibility. Time of day was not a significant factor in the baboons' selection of Opuntia fruits.

6.2 Recommendations

Recommendations for future research include a year-long study which could better distinguish between seasonality and how the troop could adapt their foraging behavior to greater changes in biomass density. Sampling females who had not become pregnant would serve as a control for the pregnant females in this study to show quantitative variation on how their energy needs changed during gestation. Timing how long each Opuntia harvesting and consumption period covered would make direct comparisons between the Opuntia fruits and the time spent on other food types possible, and the analysis more accurate.

Comparisons between length of feeding times by season and activity budgets would offer insight to the different foraging strategies among troops in the area. These suggestions could influence possible implications for management of indigenous troops.

REFERENCES

Alberts, S. C., and J. Altmann. (1995) Preparation and Activation: Determinants of Age at Reproductive Maturity in Male Baboons. *Behavioral Ecology and Sociobiology*, Vol. 36, No. 6: pp. 397-406.

Alberts S.C, and J. Altmann. (2006) The evolutionary past and the research future: environmental variation and life history flexibility in a primate lineage. In: Swedell L, Leigh SR, editors. *Reproduction and fitness in baboons: behavioral, ecological, and life history perspectives*. New York: Springer, pg. 277–303.

Alberts, S.C., Altmann, J., and M.L. Wilson. (1996) Mate guarding constrains foraging activity of male baboons. *Animal Behavior*, 51: 1269–1277.

Altmann, J. (1974). Observational Study of Behavior: Sampling Methods. *Behaviour*, 49: 227-67. http://dx.doi.org/10.1163/156853974X00534.

Altmann, J. (1980) Baboon Mothers and Infants. Cambridge, Massachusetts: Harvard University Press.

Altmann, J. & S.C. Alberts. (1987) Body mass and growth rates among wild baboons. *Oecologia* (Berl.), 72: pg. 15–20.

Altmann, J. & Muruthi, P. (1988) Differences in Daily Life Between Semiprovisioned and W i ld- Feedi ng Baboons. *American Journal of Primatology*, 15 :213- 221.

Altmann S.A, & J. Altmann. (1970) Baboon ecology: African field research. Chicago (IL): University of Chicago Press, pg. 220.

Altmann, S.A. (1974) Baboons, Space, Time and Energy. *American Zoology*, 14: 221-248.

Altmann, S.A. (1998) Foraging for survival: yearling baboons in Africa. The University of Chicago Press, Chicago.

Barton, R.A. (1989) Foraging strategies, diet and competition in olive baboons. Ph.D. Thesis,University of St. Andrews.

Barton, R.A., Whiten, A., Strum, S.C., Byrne, R.W. and A.J. Simpson. (1992) Habitat use and resource availability in baboons. *Animal Behavior*, 43: 831–844.

Barton, R.A. and A. Whiten. (1994) Reducing complex diets to simple rules: food selection by olive baboons. *Behavioural Ecology and Sociobiology,* 35: 283-293.

Basabose, A. K. (2002) Diet Composition of Chimpanzees Inhabiting the Montane Forest of Kahuzi, Democratic Republic of Congo. *American Journal of Primatology* 58:1–21.

Bercovitch, F. B. (1989) Body Size, Sperm Competition, and Determinants of Reproductive Success in Male Savanna Baboons. *Evolution*, Vol. 43, No. 7: pp. 1507-1521.

Bercovitch, F. B. and S. C. Strum. (1993) Dominance Rank, Resource Availability, and Reproductive Maturation in Female Savanna Baboons. *Behavioral Ecology and Sociobiology*, Vol. 33, No. 5: pp. 313-318.

Bio-NET-EAFRINET Keys and Fact Sheets, *Opuntia stricta*. Accessed February 2012. http://keys.lucidcentral.org/keys/v3/eafrinet/plants.htm.

Bridgeman, LeAndra Luecke. (2012). "The Feeding Ecology of Yucatán Black Howler Monkeys (*Alouatta pigra*) in Mangrove Forest, Tabasco, Mexico". Electronic Theses and Dissertations. Paper 998.

Bronikowski, A.M., and J. Altmann. (1996) Foraging in a variable environment: weather patterns and the behavioral ecology of baboons". *Behavior Ecology Sociobiology*, 39: 11–25

CABI Invasive Species Compendium online data sheet. *Opuntia stricta* (erect prickly pear). CABI Publishing 2011. www.cabi.org/ISC. Accessed February 2013.

Cant, J.G.H. and L.A. Temerin. (1984) A conceptual approach to foraging adaptation in primates. In: Adaptations for foraging in nonhuman primates: contributions to an organismal biology of Prosimians, moneys, and apes. (Eds. by Rodman, P.S. and Cant, J.G.H.) Pgs 304 - 342. Columbia University Press, New York.

Cawthon, KA. (2006) Primate Factsheets: Olive baboon (Papio anubis) Taxonomy, Morphology, & Ecology. <http://pin.primate.wisc.edu/factsheets/entry/olive_baboon>. Accessed 2011 May 20.

Charnov, E.L. (1976) Optimal foraging, the Marginal Value Theorem. *Theoretical Population Biology,* 9: 129- 136.

Corp, N. and R. W. Byrne. (2001) The Ontogeny of Manual Skill in Wild Chimpanzees: Evidence from Feeding on the Fruit of Saba Florida. *Behaviour*, 139: 137-168.

Delaney, M. J. and Happold, D.C.D. (1979) Ecology of African mammals. Longman, London.

DeVore, I. and K.R.L. Hall. (1965) Baboon ecology. In: Primate Behavior: field studies of monkeys and apes. (Ed. by DeVore, I) Pgs 20-52. Holt, Rinehart and Winston, Inc. New York.

Estes, R.D. (1991) The Behavior Guide to African Mammals. Berkley; University of California Press, pgs.509- 511.

Garber, A. (1987) Foraging Strategies among Living Primates. *Review Anthropology*, 16: 339—64.

Gaynor, D. (1994) Foraging and feeding behaviour of chacma baboons in a woodland habitat. Ph.D. Thesis, University of Natal.

Gemmill, A. and L. Gould. (2008) Microhabitat Variation and Its Effects on Dietary Composition and Intragroup Feeding Interactions Between Adult Female Lemur catta During the Dry Season at Beza Mahafaly Special Reserve, Southwestern Madagascar. *International Journal of Primatology.*

Glander, K. E., and M.F. Teaford. (1995) Seasonal, sexual, and habitat effects on feeding rates of *Alouatta* *palliata. American Journal of* *Primatology*, 36: 124–125.

Groves C. (2001) Primate taxonomy. Washington DC: Smithsonian Institute Press, pg. 350

Harding, S.R. (1973) Range utilization by a troop of Olive baboons (Papio anubis). Ph.D. thesis, University of California, Berkeley.

Heller, J.A., Knott, C.D., Conklin-Brittain, N.L., Rudel, L.L., Wilson, M.D. and J.W. Froehlich. (2002) Fatty acid profiles of orangutan (Pongo pygmaeus) foods as determined by gas-liquid chromatography: cambium, seeds and fruit. *American Journal of Primatology*, 57(1):44.

Hill, R.A., and R.I.M. Dunbar. (2002) Climatic determinants of diet and foraging behavior in baboons. *Evolutionary Ecology* 16: 579–593.

ILWIS .3.7.2. 2007. <www.ilwis.org>

Jaman, M.F., Huffman, F.A., and H. Takemoto. (2010). The foraging behavior of Japanese macaques (*Macaca fuscata*) in a forested enclosure: Effects of nutrient composition, energy and its seasonal variation on the consumption of natural plant foods. *Current Zoology*, 56 (2): 198-208.

Janson, C. H., and C.A. Chapman. (1999) Resources and primate community structure. In J. G. Fleagle, C. H. Janson, & K. E. Reed (Eds.), Primate communities (pp. 237–267). Cambridge, UK: Cambridge University Press.

Kunyanga, C. and J.K. Imungi. (2009) Prickly cactus pear fruit (Opuntia spp). Department of Food Science, Nutrition and Technology, Faculty of Agriculture, College of Agriculture and Veterinary Sciences, University of Nairobi, P.O. Box 29053-00625, Nairobi.

Kunz, B.K. and K.E. Linsenmair. (2007) Changes in baboon feeding behaviour: maturity-dependent fruit and seed size selection within a food plant species. *International Journal of Primatology* 28: 819-835.

Laikipia Wildlife Forum. "Rangeland Rehabilitation". Laikipia Tourism, 2013. Accessed 27 June 2013. <http://www.laikipia.org/programmes-top/rangeland-rehabilitation>

McFarland, D. (1994) *Animal behavior*. John Wiley and Sons Inc., pp 152-161.

McNaughton, S. J. (1979) Grassland-Herbivore Dynamics, pp 46-50. In: Serengeti dynamics of an ecosystem. (Eds. Sinclair, A.R.E. and M. Norton Griffiths). University of Chicago Press.

MacArthur, R. H. and E.R. Pianka. (1966) On the optimal use of a patchy environment. *American Naturalist*, pg. 100.

Mellgren, R.L. and S.W. Brown. (1987) Environmental constraints on optimal- foraging behavior. In: Quantitative analysis of behavior: Foraging (Ed. by Commons, M.L., Kacelnik, A. and Shettleworth, S.J.). Pgs 133-151. Lawrence Erlbaum Associates, Publishers New Jersey.

Muruthi, P., Altmann, J., and S. Altmann. (1991) Resource base, parity, and reproductive condition affect females' feeding time and nutrient intake within and between groups of a baboon population. *Oecologia*, 87: 467 – 472.

Mutua, C. M. (2001) The Role of Topography on Feeding and Ranging Patterns of Olive Baboons (Papio anubis) in a dry Savanna Environment. Moi University, Eldoret, M. Phil.

Natural Resources Conservation Service (NRCS), US Department of Agriculture. Plants Database, Plants Profile for *Opuntia stricta*. Accessed 15 June 2011. <http://plants.usda.gov/java/profile?symbol=OPFI.>

Okecha, A.A. and N.E. Newton-Fisher. (2006) The Diet of Olive Baboons (Papio Anubis) in the Budongo Forest Reserve, Uganda. SVNY253-Newton-Fisher *et al.* July 11, 0:31.

Post, D.G. (1982) Feeding behavior of yellow baboons (Papio cynocephalus) in the Amboseli National Park, Kenya. *International Journal Primatology*, 3(4): 403-30.

Sayers, K., Norconk, M.A., & N. L. Conklin-Brittain. (2009) Optimal Foraging on the Roof of the World: Himalayan Langurs and the Classical Prey Model. *American Journal of Physical Anthropology,* 000:000–000, DOI 10.1002/ajpa.21149.

Segal, C. (2008) Foraging Behavior and Diet in Chacma Baboons in Suikerbosrand Nature Reserve. University of the Witwatersrand, Johannesburn, M.Sc.

Silk, J. B. (2002) Practice Random Acts of Aggression and Senseless Acts of Intimidation: The Logic of Status Contests in Social Groups. *Evolutionary Anthropology,* 11:221–225 DOI 10.1002/evan.10038

SPSS. 16.0. (2007) SPSS Inc. <http://spss.en.softonic.com>

Strum, S. C. (1975) Primate Predation: Interim Report on the Development of a Tradition in a Troop of Olive Baboons. *Science,* New Series, Vol. 187, No. 4178: 755-757. http://www.jstor.org/stable/1739822.

Strum, S. C. (1982) Agnostic Dominance in Male Baboons: An alternative view. *International Journal of* *Primatology*, Vol. 3, No. 2.

Strum, S. C. (1991) Weight and age in wild olive baboons. *American Journal of Primatology*, 25: 219–237.

Strum, S. C. (2005) Measuring Success in Primate Translocation: A Baboon Case Study. *American Journal of Primatology,* 65:117–140.

Strum, S. C. (2009) The Development of Primate Raiding: implications for management and conservation. *Int J Primatol*, 31(1): 133–156.   Published online 2010 January 8. doi: 10.1007/s10764-009-9387-5

Survey Branch of the Kenya Forest Department. "Mukogodo Forest Reserve". Protected Planet, 2012. Accessed   27   June   2013. <http://www.protectedplanet.net/sites/Mukogodo_Forest_Reserve.>

Sussman RW. (2000). Primate ecology and social structure. Volume 2: New World monkeys. Heights, MA: Pearson Custom Publishing.

Swedell, L., Hailemeskel, G., and A. Schreier. (2008) Composition and Seasonality of Diet in Wild Hamadryas   Baboons: Preliminary Findings from Filoha. *Folia Primatology*, 79:476–490.

Wahungu, G. (1998) Diet and habitat overlap in two sympatric primate species,   the Tana crested   mangabey Cercocebus galeritus and yellow baboon Papio cynocephalus. *African Journal of Ecology*, 36:   159–173. doi: 10.1046/j.1365-2028.1998.00120.

Walker, W.E., Marshell, S.D., Rypstra, A.L. and D.H. Taylor. (1999) The effect of hunger on locomotary behaviour in two species of wolf spider (Araneae, Lycosidae). *Animal Behaviour*, 58: 515-520.

Watts, D. P. and J. C., Mitani. (2000) Hunting Behavior of Chimpanzees at Ngogo, Kibale National Park, Uganda. *Behaviour*, 138: 299-327.

Whiten, A., Byrne, R.W., Barton, R.A., Waterman, P.G., Henzi, S.P., Hawkes, K., Widowson, E.M., Altmann, S.A., Milton, K. and R.I.M. Dunbar. (1991) Dietary and foraging strategies of baboons. Philosophical Transactions: *Biological Sciences*, 334(1270): 187- 197.

Whitten, P. L. (1982) Female reproductive strategies among vervet monkeys. Ph.D. dissertation, Harvard University.

APPENDIX

Appendix 1: Frequency of follows per individual, with total length of observations in hours, and the dates for each follow are included.

Individual (ID) animal	Morning follows, dd/mm/yr	Full day follows, dd/mm/yr	Length of observation, hrs
H1 (m)	25/8/11		6:00:00
	14/9/11		6:00:00
		5/10/11	10:00:00
	17/10/11		5:50:00
	31/10/11		6:00:00
	23/11/11		6:00:00
	14/1/12		6:00:00
		28/1/12	9:55:00
KT (f)		8/9/11	10:00:00
	21/9/11		6:00:00
	6/10/11		6:00:00
	16/10/11		6:00:00
	2/11/11		5:58:00
	21/11/11		6:00:00
	17/1/12		6:00:00
	26/1/12		6:00:00
NP (m)	6/9/11		6:00:00
	17/9/11		6:00:00
	30/9/11		6:00:00
		13/10/11	9:50:00
	14/11/11		6:00:00
	29/11/11		6:00:00
	21/1/12		6:00:00
		22/2/12	9:40:00
PU (m)	23/8/11		6:00:00
	26/8/11		6:00:00
		22/9/11	10:00:00

	11/10/11		6:00:00
	24/10/11		6:00:00
	22/11/11		5:55:00
	23/2/12		6:00:00
SB (f)	31/8/11		6:00:00
	13/9/11		6:00:00
		29/9/11	10:00:00
	10/10/11		5:50:00
	26/10/11		6:00:00
	15/11/11		6:00:00
	16/1/12		6:00:00
	30/1/12		6:00:00
TL (f)	22/8/11		6:00:00
	10/9/11		6:00:00
	28/9/11		6:00:00
	18/10/11		6:00:00
	10/11/11		6:00:00
	18/11/11		6:00:00
		2/12/11	9:08:00
	25/1/12		6:00:00
TO (f)	5/9/11		6:00:00
		15/9/11	10:00:00
	1/10/11		5:50:00
	15/10/11		6:00:00
	7/11/11		5:20:00
	11/11/11		5:25:00
	13/1/12		6:00:00
	27/1/12		6:00:00
XV (m)	3/9/11		5:20:00
	27/9/11		6:00:00
	4/10/11		6:00:00
	27/10/11		6:00:00

		3/11/11	9:30:00
	19/11/11		6:00:00
	22/1/12		6:00:00
	16/2/12		6:00:00
Totals	**53**	**10**	**916**

(NOTE: (m) = males; (f) = females

Appendix 2: UNBP Food List

UNBP Baboon Food List 2012

A GRASSES, SEDGES
1 Grass. no spp.
2 Cynodon (both spp.)
3 Tragus bertorianus
4 Cyperus spp
5 Eragrostis spp.
6 Kyllinga allata* (yellow flower)
7 Kyllinga alba* (white flower)
8 Sporobulus spp.
90 Mariscus amauropeus
91 Dactyloctenium aegyptium
92 Penissetum (both spp.)
95 Species seen/not coded above**

B HERBS
11 Herb. no spp.
12 Conostomium quadrangulare
13 Amaranthus graecizans
14 Tribulus terrestris
15 Hibiscus spp.
16 Oxygonum sinuatum
17 Plectranthus caninus =
18 Asparagus spp.
19 Monsonia augustifolia
20 Mexican marigold (Tagetas minuta)
21 Commelina spp.
22 Polichia compestris
24 Erucastrum arabican =
26 Ipomoea kituiensis (big) *
27 Osteospermum vaillantii
28 Trachyandra saltii
29 Ipomoea mombasana (little)
30 Jasminum fluminense *
36 Saprophyte **
38 Becium spp. *
40 Delosperma nakurense +
58 Talinum crispatulatum +
59 Polygala sphenoptera +
60 Lactuca capensis +
63 Coccinia adoensis (melon)
84 Albuca wakefieldii +
86 Ammocharis spp. (small/large) *
88 Lippia ^
89 Herbs, mixed handful *
96 Spp seen/not coded above **
111 Commicapus pedunculosus +
115 Portulaca oleracea +
116 Indigofera ^
140 Cyphostemma orondo ~
141 Peponium vogelii ~
150 Craterostigma++++
200 Feeding on the ground. no spp. xxx
201 Feeding underground. no spp. xxx

Special Notes:

(a) Number 121 was first eaten in 1995-96 but coded as 32; new code added 3/1/99
(b) Acacis seedpod codes comprehensively used from 09/06
(c) Opuntia # 121 fruit codes added 03/07
(d) Species name for Opuntia 2 (number 121) added 1/10
(e) Number 114 reclassified from herb to bush 1/10
(f) H and SH used interchangeably for seedhead until 1/10
(g) GSP and GS used interchangeably until 2/10. GP added 2/10 and seedpod codes clarified

C SUCCULENTS
9 Sanseveria abyssinica (large)
10 Sanseveria intermedia (small)
23 Euphorbia heterochroma (small)
32 Opuntia vulgaris # 1
34 Aloe (secundiflora)
35 Agave sisalana
93 Cissus rotundifolia*
94 Euphorbia gossypina*
120 Opuntia # 3 ^^
121 Opuntia stricta # 2 ^^
122 Opuntia # 4 ++
124 Carraluma++++

D BUSHES
25 Solanum
31 Bush. No spp.
33 Croton dichogamus
37 Carissa edulis
39 Pyristria phyllanthoidea +
53 Lycium europaeum
97 Spp seen/not coded above
99 New food, not seen before (list part)
113 Solanum nigrum ^^
114 Pavetta gardenlifolia ^
117 Melhania velutine ^^
151 Grewia tebensis ♦

E PLANT PART CODES
81 Seeds no spp
L Leaves
S Seeds
SP Seed Pods
F Flowers
B Bark
R Fruit
E Exudate
TH Thorn
G Grass blade

TT	Stem		
FB	Flowerbud		
RO	Root		
C	Corm		
IG	Insectgall		
SB	Stembase		
H	Seedhead		

Opuntia ~ 121 Fruit Codes
RGR Green Fruit
RTR Transitional Fruit
RR1 Ripe Fruit
RR2 Very Ripe Fruit (dark colour)
RDS Discard
RDR Dried Fruit
ES Seeds from Elephant dung ♦
DRD Dry Discard ♦
NP New Pad ♦

Acacia Seedpod Codes:
GSP Seeds from green pod
DSP Seeds from dry pod
DS Dry seeds on ground
GS Green seeds on ground
GP Green pod ■

F TREES
41 Tree no spp.
42 Ficus spp.
43 Grewia bicolar
44 Rhus natalensis
45 Acacia etbaica
46 Acacia drepanolobium
47 Acacia xanthophloea
48 Acacia tortilis
49 Euphorbia kibwezensis =
50 Cordia spp.
51 Commiphora schimperi =
52 A. Mellifera (wait-a-bit)
54 Acacia nilotica
55 Boscia anguistifolia =
56 Acacia brevispica
57 Dracaena ellenbeckiana *
85 Acacia seyal ++
98 Balanites aegyptiaca ►
112 Papea capensis^
118 Teclea nobilis ^^
119 Haplocococelum foliolosum ^^
123 Acacia ++
130 Commiphora (sp 2)++++
131 Euclea divinorum ~
132 Acokanthera schimperi ~

G MISC
61 Fruit. no spp.
62 Bud. no spp.
64 Insect, no spp.
65 Larvae, no spp.
66 Bird, no spp.
67 Egg, no spp.
68 Salt Lick Gully
69 Caterpillar, no spp. *
70 Snails, no spp.
71 Mushrooms, no spp.
72 Feces
73 Human food
74 Mammal, no spp.
75 Scorpion *
76 Insects, under rocks *
77 Grasshopper *
78 Termite *
79 Cocoon *
80 Cape hare *
82 Dikdik **
83 Gazelle **
87 Guinea fowl young *
100 Lizard +++
102 Tortoise++++
103 Ant Mound Debris ■

*	New 1993	(n = 18)
**	New 1994	(n = 5)
+	New 1995	(n = 8)
=	Renamed 1995	(n = 5)
^	New 1997	(n = 4)
^^	New 1998	(n = 6)
►	New 2000	(n = 1)
++	New 2001(July)	(n = 3)
+++	New 2002	(n = 1)
++++	New 2003	(n = 4)
xxx	New 2006	(n = 2)
~	New 2008	(n = 4)
♦	New 2009	(n = 5)
■	New 2010/11	(n = 2)

Appendix 3: Activity Codes (UNBP Protocol)

K – walking and foraging

F – stationary feeding or harvesting on one food species

T – movement at a faster pace than normal without foraging or feeding, to include times when a fast pace is used as when targeting a specific location or avoiding external interference

R – any type of resting for long periods

S – any type of social activity (ie, aggression, greetings, male-infant, attention to infants, avoids)

G – Individual (ID) is grooming another ID/ID is being groomed

Appendix 4: Plant species eaten during August 2011 – February 2012 that contributed to at least 1% of total feeding time for all individuals followed with total time spent on each species, the part eaten, and the class of food type. Each month was assessed individually.

Month	Species	Food Type	Part Eaten	Total Feeding Time per month, min
August 2011	*Tribulus terrestris*	grass/herb	leaves, stem	187.50
	Cynodon	grass/herb	grass blade, seed head	104.40
	Pyristria phyllanthoidea	bush/tree	fruit	21.00
	Lycium europaeum	bush/tree	leaves	9.10
	Acacia mellifera	bush/tree	leaves, bark	9.10

	Tragus bertorianus	grass/herb	grass blade, seed head	7.20
	Drinking, salt lick	misc.		6.30
	Kyllinga allata	grass/herb	grass blade	4.50
September 2011	*Tribulus terrestris*	grass/herb	leaves, fruit, stem base	476.10
	Tragus bertorianus	grass/herb	grass blade, seed head	136.10
	Cynodon	grass/herb	grass blade, stem base, seed head	87.50
	Ipomoea mombasana	grass/herb	flower, fruit	47.00
	Drinking, salt lick	misc.		43.70
	Commelina	grass/herb	leaves, stem base	42.20
	Sanseveria intermedia	succulent	stem, stem base	35.30
	Lycium europaeum	bush/tree	leaves	32.00
	Pyristria phyllanthoidea	bush/tree	fruit	29.00
	Cyperus	grass/herb	stem base, grass blade	26.20
	Herb, no spp	grass/herb	leaves	19.20
	Conostomium quadrangulare	grass/herb	stem	17.50
	Ammocharis	grass/herb	leaves	15.50
	Osteospermum vaillantii	grass/herb	leaves	14.00
	Opuntia spp, other	succulent	pads	12.30
October 2011	*Cynodon*	grass/herb	grass blade, stem base	172.30

	Penissetum	grass/herb	grass blade	102.10
	Tragus bertorianus	grass/herb	grass blade	73.30
	Sanseveria intermedia	succulent	stem, stem base	71.40
	Lycium europaeum	bush/tree	leaves, fruit	59.40
	Tribulus terrestris	grass/herb	fruit, leaves	57.00
	Predation (dik dik, duck chicks, cape hare)	misc.		43.70
	Ipomoea mombasana	grass/herb	flower bud	36.20
	Acacia etbaica	bush/tree	flower, leaves	31.20
	Feeding on ground, no spp	misc.		29.70
	Pyristria phyllanthoidea	bush/tree	fruit	28.00
	Teclea nobilis	bush/tree	fruit	28.00
	Acacia mellifera	bush/tree	leaves, flower, bark	26.40
	Acacia xanthophloea	bush/tree	exudate, bark	18.10
	Commelina	grass/herb	leaves, stem base	13.50
	Asparagus	grass/herb	leaves, stem	13.00
	Erucastrum arabican	grass/herb	flower	12.30
	Bush, no spp	bush/tree	leaves, fruit	12.20
November 2011	*Penissetum*	grass/herb	grass blade	111.20
	Tragus bertorianus	grass/herb	grass blade	88.50
	Cynodon	grass/herb	grass blade	64.20
	Ammocharis	grass/herb	leaves	46.10

	Lycium europaeum	bush/tree	leaves	43.50
	Asparagus	grass/herb	leaves	27.00
	Polichia compestris	grass/herb	fruit	27.00
	Commelina	grass/herb	leaf base, flower bud	26.50
	Tribulus terrestris	grass/herb	leaves, fruit	26.30
	Drinking, salt lick	misc.		26.20
	Eragrostis	grass/herb	stem base, grass blade	23.50
	Ipomoea mombasana	grass/herb	flower, flower bud	16.40
	Cyperus	grass/herb	grass blade, stem base	16.40
	Grewia bicolar	bush/tree	leaves	15.10
	Oxygonum sinuatum	grass/herb	leaves	14.30
	Sanseveria intermedia	succulent	stem, root, fruit	14.10
	Lippia	grass/herb	leaves	13.10
	Ant mound debris	misc.		12.90
	Acacia etbaica	bush/tree	bark, leaves	9.30
	Acacia mellifera	bush/tree	leaves, bark, flower	9.10
	Invertebrates	misc.		8.50
	Albuca wakefieldii	grass/herb	root, leaves	8.10
	Mushrooms	misc.		8.00
	Herb, no spp	grass/herb	leaves, new stem	7.50
January 2012	*Cynodon*	grass/herb	corm, stem base, grass blade	255.00

	Ipomoea mombasana	grass/herb	flower bud, flower	109.10
	Tragus bertorianus	grass/herb	corm, stem base, grass blade	56.00
	Polichia compestris	grass/herb	fruit	45.30
	Sanseveria intermedia	succulent	stem, stem base, root	42.30
	Feeding on ground	misc.		32.40
	Penissetum	grass/herb	stem base, grass blade, corm	25.50
	Invertebrates	misc.		25.10
	Lycium europaeum	bush/tree	leaves	21.30
	Lippia	grass/herb	stem base, flower bud	18.20
	Amaranthus graecizans	grass/herb	flower bud, leaves	15.10
	Commelina	grass/herb	leaf base	13.20
	Grewia bicolar	bush/tree	leaves	12.20
	Conostomium quadrangulare	grass/herb	stem, stem base	8.50
	Solanum	bush/tree	flower bud, fruit	8.50
	Predation (egg, Guinea fowl young, Cape Hare)	misc.		8.10
February 2012	*Acacia tortilis*	bush/tree	flower	93.50
	Acacia etbaica	bush/tree	flower	74.10
	Cynodon	grass/herb	grass blade, stem base	35.20

Cyperus	grass/herb	corm	34.40
Polichia compestris	grass/herb	fruit	15.00
Penissetum	grass/herb	grass blade	7.10
Grass, no spp	grass/herb	grass blade	6.10

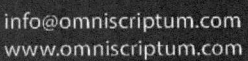